U0159227

单板滑雪

SNOWBOARDING

单板滑雪
SNOWBOARDING

　　这可是魅力四射的运动。踩着滑板，在规定的山坡线路上回转滑雪，或在特设的场地内旋转起跳，还要做各种炫酷的高难度动作。

Drop downhill the curves on a snowboard. Or perform complicated moves off a half-pipe. Different venues but the same fascination.

BIATHLON

冬季两项
BIATHLON

　　运动员们都是"飞行军"，身背长枪，脚踏滑雪板，在10～20公里的越野赛道追逐竞速，同时也是"神枪手"，一边飞驰一边打靶。

　　Racing through a 10-20 kilometer cross-country course, we are the Flashes on skiing boards! Precisely shooting at each target along the track, we are the Hawkeyes with rifles!

跳台滑雪

SKI JUMPING

跳台滑雪
SKI JUMPING

从长长的坡道俯冲而下，从跳台上一跃飞出，在空中展示优美动作，比的是谁跳得又远又好看，实在太刺激了！

We dash down the track. We jump off the ramp. We fly in the air! To be the one reaching the furthest! To be the one moving the most elegantly! Hold your breath, never blink.

花样滑水

FIGURE SKATING

花样滑冰
FIGURE SKATING

这可是最受欢迎的冰上运动项目，是艺术与运动完美结合的典范！伴随着优美的音乐，运动员在冰上优雅滑行，翩翩起舞，简直就是冰上精灵！

The most popular ice sport. The perfect combination of art and sport. To the melodies, we glide and we dance, like fairies on the ice.

SPEED SKATING

速度滑冰
SPEED SKATING

　　踩着冰刀，蹬冰、收腿、下刀，全速冲刺！是不是像看武功高手过招一样过瘾？天下武功，唯快不破。

Stand on blades, take a run-up, draw back the leg, push off the ice and dash forward! Doesn't it look like a Wulin tournament? Diamond cut diamond and the fastest will be crowned.

冰
壶

CURLING

冰 壶
CURLING

被称为"冰上国际象棋"的冰壶，不仅要掷得远，更要掷得准，不仅自己要占据有利位置，还要想方设法把对方打出去。

It's titled as the chess on the ice, requiring both strength and accuracy. We should guard our vantage point and hit rivals' curling stones out the house.

冰球

ICE HOCKEY

冰 球
ICE HOCKEY

就是冰上玩的曲棍球嘛，运动员们脚蹬冰刀，手持球棍，奋力冲击，激烈对抗。据说每名运动员上场拼杀 30 ~ 120 秒就必须轮换休息。

Skate on the ice and use long sticks to hit a hard rubber disc into the other team's goal. We chase, and we fight. No way for keeping playing in the rink for more than 2 minutes.

越野滑雪

CROSS-COUNTRY SKIING

越野滑雪
CROSS-COUNTRY SKIING

这是世界上最古老的运动项目之一。踏着滑雪板，穿梭在山林湖泊之间，乐哉美哉！其实非常考验运动员的技术和经验。

One of the oldest sports in the world. Ski across the snow-covered countryside and travel through the nature. Sounds interesting, isn't it? But it demands great skills and rich experience.

SKELETON

钢架雪车
SKELETON

　　运动员在雪橇上保持头朝前的俯身姿势，靠肩膀和膝盖来控制方向，在1200m以上的倾斜冰道上全速滑行。注意，钢架雪车，弯道不减速哦!

Lie prostrate and face forward please, we will dash down a 1200-meter ice track on a skeleton. Our knees and arms are the only steering wheels. Warm prompt: no speed reduction when passing the curves.

BOBSLEIGHING AND TOBOGGANING

雪 车
BOBSLEIGHING AND TOBOGGANING

这不就是在冰道上玩"碰碰车"吗？雪车，也叫"有舵雪橇"，集体乘坐，前有"司机"操纵舵把控制方向，后有"司闸员"负责制动。冲啊！我们一起飙雪车吧！

Also known as bobsled, it's a team sport of ski racing on a bobsleigh which looks so like a bumper car. We have drivers steering the sleds, and brakemen in the rear. The F1 on the ice is beginning！ Are you ready? Let's go!

FREESTYLE SKIING

自由式滑雪
FREESTYLE SKIING

滑雪能有多么自由？运动员们在专门的滑雪场上，完成空中技巧、雪上技巧和雪上芭蕾的视觉盛宴。看上去，是不是有点像雪地特技？

How free can skiing be? In a special ski field, you can enjoy a visual feast of aerials, moguls, and acroski. Wait a minute, isn't that the acrobatics above snowfield?

ALPINE SKIING

高山滑雪
ALPINE SKIING

又叫"阿尔卑斯滑雪"，由越野滑雪衍生而来。运动员们在设定的赛道上，从高山俯冲而下。比得就是速度！

A sport deriving from cross-country skiing. Along a special track, we slide down snow-covered hills. Nothing matters here, and only speed talks.

雪橇

LUGE

雪橇
LUGE

　　一名或两名运动员，背贴雪橇仰卧，双脚在前，通过身体控制方向，高速回转滑降。如果在这个项目上拿到奖牌，就真正实现了"躺赢"。

　　One or two players ride down a track of ice, back against the luge and feet down. If we can get a medal in this game, that is a real lying win.